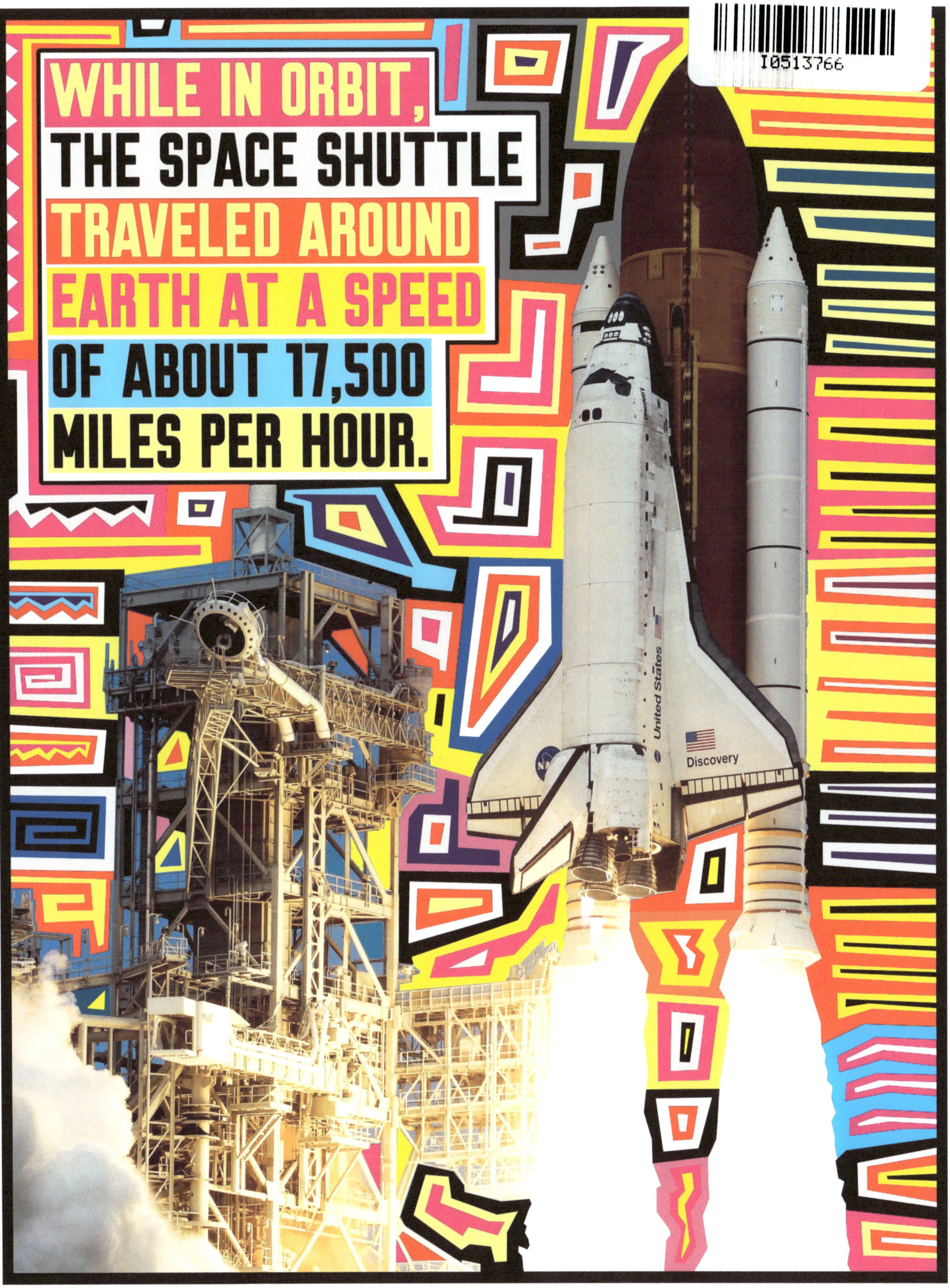
WHILE IN ORBIT,
THE SPACE SHUTTLE
TRAVELED AROUND
EARTH AT A SPEED
OF ABOUT 17,500
MILES PER HOUR.
United States
Discovery

NASA

NASA BEGAN TO OPERATE IN 1958, ONE YEAR AFTER SPUTNIK 1 WAS LAUNCHED.

A PERSON CAN'T BE RECOGNIZED AS AN ASTRONAUT
BY NASA UNTIL THEY HAVE TRAVELED 50 MILES
FROM THE SURFACE OF THE EARTH.

THE SPACE SHUTTLE
COMPLETED 135 MISSIONS.

THE HUBBLE SPACE TELESCOPE

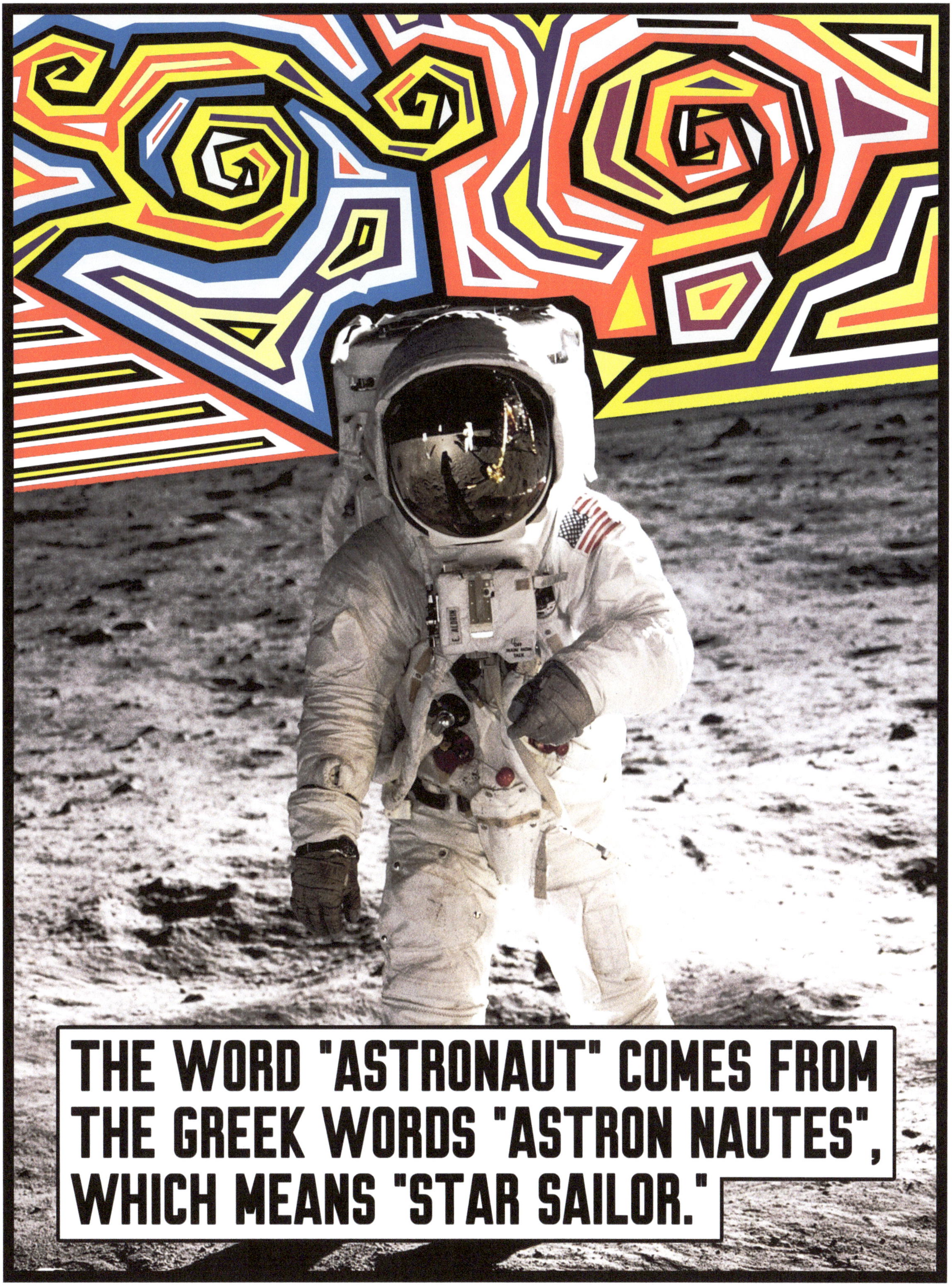
THE WORD "ASTRONAUT" COMES FROM THE GREEK WORDS "ASTRON NAUTES", WHICH MEANS "STAR SAILOR."

ASTRONAUT
WORKING ON THE
HUBBLE SPACE
TELESCOPE.

NEIL ARMSTRONG
WAS THE FIRST
PERSON TO WALK
ON THE MOON.
NASA
ARMSTRONG

BUZZ ALDRIN
WAS THE SECOND
PERSON TO WALK
ON THE MOON.
NASA
ALDRIN

Discovery
United States
SPACE SHUTTLE DISCOVERY

PUTTING A SPACESUIT ON IS CALLED "DONNING" THE SUIT. REMOVING THE SUIT IS CALLED "DOFFING."
NASA

THANK YOU.
THE END.

SPACE
IMAGES
COURTESY
OF NASA.

631ART.COM PRESENTS:
MAGIC CAT KINGDOM
BY EDDIE ALFARO
the town of DISTOPIA
ANTI-SOCIAL MEDIA
BY EDDIE ALFARO
631ART.COM PRESENTS:
BLISSFUL
A COLORING BOOK.
BY EDDIE ALFARO
631ART.COM PRESENTS:
INTERESTING FACTS ABOUT
HISTORICAL FIGURES
BY EDDIE ALFARO
MORE BOOKS AT:
631ART.COM PRESENTS:
THE BOOK OF TIGERS
BY EDDIE ALFARO
BUDDAKATS
"THE BEAUTIFUL WORLD"
BY EDDIE ALFARO
631ART.COM PRESENTS:
SNUGS FOR THE PUGS
BY EDDIE ALFARO
631ART.COM PRESENTS:
INTERESTING FACTS OF
GREEK MYTHOLOGY
BY EDDIE ALFARO
631ART.COM
631ART.COM PRESENTS:
ROTTWEILERS
BY EDDIE ALFARO
VALLEY OF the HYPNOTIZED
"YOU STRESS ME OUT"
BY EDDIE ALFARO
631ART.COM PRESENTS:
INTERESTING FACTS ABOUT
ABRAHAM LINCOLN
BY EDDIE ALFARO
631ART.COM PRESENTS:
"ANCIENT EGYPT"
A COLORING BOOK
BY EDDIE ALFARO

www.ingramcontent.com/pod-product-compliance
Lightning Source LLC
Chambersburg PA
CBHW040407220526
45473CB00004B/1159